Saving Our Planet

Saving Earth's Animals

by Brienna Rossiter

FOCUS READERS®

PIONEER

www.focusreaders.com

Focus Readers is distributed by North Star Editions:
sales@northstareditions.com | 888-417-0195

Produced for Focus Readers by Red Line Editorial.

Photographs ©: Shutterstock Images, cover, 1, 4, 8, 11, 12, 15, 16, 19 (top), 19 (bottom), 21; Red Line Editorial, 7

Library of Congress Cataloging-in-Publication Data
Names: Rossiter, Brienna, author.
Title: Saving earth's animals / by Brienna Rossiter.
Description: Lake Elmo, MN : Focus Readers, [2022] | Series: Saving our planet | Includes index. | Audience: Grades 2-3
Identifiers: LCCN 2021006323 (print) | LCCN 2021006324 (ebook) | ISBN 9781644938225 (hardcover) | ISBN 9781644938683 (paperback) | ISBN 9781644939147 (ebook) | ISBN 9781644939598 (pdf)
Subjects: LCSH: Endangered species--Conservation--Juvenile literature. | Wildlife conservation--Juvenile literature.
Classification: LCC QL83 .R67 2022 (print) | LCC QL83 (ebook) | DDC 591.68--dc23
LC record available at https://lccn.loc.gov/2021006323
LC ebook record available at https://lccn.loc.gov/2021006324

Printed in the United States of America
Mankato, MN
082021

About the Author

Brienna Rossiter is a writer and editor who lives in Minnesota. She loves looking for animals while going on hikes.

Table of Contents

Chapter 1

Endangered Animals

Many different animals live on Earth. But some of them face **threats**. Their **populations** are getting smaller. The places they live are shrinking, too.

Some animals do not have enough food. Others do not have enough space to live. As a result, these animals are at risk of dying out. One day, they may be gone completely.

Threat Categories

Scientists use categories to show which animals are most at risk.

Extinct	The animal lives nowhere on Earth.
Extinct in the Wild	The animal is no longer alive in the wild.
Critically Endangered	The animal's population is very small or shrinking very quickly.
Endangered	The animal's population is small or shrinking quickly.
Vulnerable	The animal's population is dropping enough to be at risk.
Near Threatened	The animal may be at risk soon.
Least Concern	The animal is not at risk.

Risk of Extinction

Chapter 2

Causes

Some animals die out because of **predators**. Others die out because of disease. But the biggest threats come from humans. People may hunt or fish too much.

Humans also harm animal **habitats**. People clear land. They make buildings or farms. They build roads. They bring in new plants and animals. People also **pollute**. These changes make it hard for animals to survive.

Chapter 3

Impacts

People move into new areas. Animals are forced out. The animals may not be able to find food or **shelter**. They may not be able to raise babies.

Changes to one animal can affect many others. For example, many animals eat insects. If insects disappear, other animals will not have food.

Did You Know?

Sometimes people bring a **species** to a new place. It can hurt plants or animals that already live there.

Chapter 4

What We Can Do

People can work to protect Earth's animals. Governments can make laws against hunting rare animals. And they can set aside safe places for animals to live.

Conservation groups help, too. These groups study where animals live. They study what threats animals face. They find ways to save the animals. And they help people use land in less harmful ways.

Did You Know?

Some groups help raise animals. Then they release the animals back into the wild.

BEACHFRONT
LIGHTS OUT
MAY–OCTOBER
Let's Save
The Sea Turtles
Sponsored by:
U.S. Fish and Wildlife Service
S.C. Department of Natural Resources
SCE&G

Healthy Habitats

You can help animals by protecting their habitats. When you walk or hike, stay on paths. Don't take plants or animals home. And don't leave behind trash. Trash can make animals sick or hurt. By keeping nature clean, you can keep habitats safe.

FOCUS ON

Saving Earth's Animals

Write your answers on a separate piece of paper.

1. Write a sentence describing the main ideas of Chapter 1.
2. Do you see many wild animals near your home or school? Why do you think that is?
3. What is the main threat to most animals?
   A. predators
   B. humans
   C. disease
4. If all the rabbits in an area die, what might happen to other animals in that area?
   A. Animals that eat rabbits might die, too.
   B. The other animals would all be healthier.
   C. The other animals would not be affected.

Answer key on page 24.

Glossary

conservation
Working to protect plants, animals, and natural resources so they are not lost or wasted.

habitats
Places where animals live.

pollute
To release harmful chemicals into water, air, or soil.

populations
The numbers of animals living in certain areas.

predators
Animals that hunt other animals for food.

shelter
A place to live or hide.

species
A group of animals or plants of the same kind.

threats
Things that cause problems or danger.

To Learn More

BOOKS

Blakemore, Megan. *The Threatened Lives of Chimpanzees*. Mankato, MN: The Child's World, 2018.

Kukla, Lauren. *Protecting Ocean Animals*. Minneapolis: Abdo Publishing, 2017.

NOTE TO EDUCATORS

Visit **www.focusreaders.com** to find lesson plans, activities, links, and other resources related to this title.

Index

Answer Key: 1. Answers will vary; **2.** Answers will vary; **3.** B; **4.** A